Erosion

Erosion

written and photographed by
Cherie Winner

A Carolrhoda Earth Watch Book

Carolrhoda Books, Inc./Minneapolis

*For David and Jane Love, for their gracious-
ness, good humor, and love of the land*

The author thanks Dr. Carrick Eggleston,
Department of Geology, University of Wyoming;
Dr. Alan Cvancara; and Dr. Jerry Nelson,
Department of Geology, Casper College, for
their time and patience in explaining complex
concepts; and Tom Morton, Vernon and Helen
Olsen, and Leila Stanfield for helping me get
photos of worms, arches, and the Big Hollow.

Carolrhoda Books, Inc., c/o The Lerner Publishing Group
241 First Avenue North, Minneapolis, MN 55401 U.S.A.

Website address: www.lernerbooks.com

Library of Congress Cataloging-in-Publication Data

Winner, Cherie.
   Erosion / written and photographed by Cherie Winner.
    p. cm.
   "A Carolrhoda earth watch book."
   Includes index.
   Summary: Describes the forces of erosion as caused by
glaciers, water, and wind, how they affect the earth's sur-
face, and how their destructive effects can be prevented.
   ISBN 1-57505-223-7 (alk. paper)
   1. Erosion—Juvenile literature. [1. Erosion.] I. Title.
QE571.W55 1999
551.3'02—dc21   98-16456

Manufactured in the United States of America
  2 3 4 5 6 – JR – 04 03 02

CONTENTS

THE CHANGING EARTH

The earth changes all the time. Some forces of change, such as volcanoes and earthquakes, are violent. A volcano like Mount St. Helens, which erupted in Washington State in 1980, can bury cars, towns, and forests in a blizzard of ash. When a big earthquake hits, buildings and roads crumble. Huge gaps can be ripped into the earth within a few minutes.

But not all forces produce changes that are as sudden and spectacular as these. The process called **erosion** makes huge changes in the landscape, but it works slowly, steadily, and quietly. Moving a few grains of sand at a time, rivers carve deep paths into the earth. Ice, wind, and rain soften the jagged outlines of mountain peaks. Sand carried by the wind sculpts weird shapes out of desert rock.

Unlike volcanoes and earthquakes, erosion doesn't happen only every few years. It's happening right now. All over the world, every day, every minute, erosion is changing the face of the earth.

WHAT IS EROSION?

The word *erosion* comes from the ancient Latin word *erodere*, which means "to slowly eat away." Erosion is the process that loosens rock, dirt, and other bits of earth in one place and moves them to another place. During erosion, water, wind, and huge masses of ice called **glaciers**—all aided by gravity—break off pieces of earth, carry them away, and drop, or **deposit,** them somewhere else.

The first step in erosion is the formation of loose bits of earth. This is called **weathering.** Some of the loose material that results from weathering is carried away by wind, water, or ice. Some of it remains in place and becomes soil. Soil is more than just dirt. Living things, or organisms, are contained within it. Organisms in the soil include earthworms, fungi (such as mushrooms), and bacteria, which are tiny microorganisms too small for a person to see. Soil also contains the remains of living things, like dead plant roots and dead leaves.

Unlike erosion, earthquakes and volcanoes can cause drastic changes to the landscape in very short periods of time. An earthquake ripped huge gaps into the earth in Guatemala (left), and the blast from Mount St. Helens leveled an entire forest (right).

Trees and shrubs sprout from cracks in a large rock face, causing the rock to break apart. Some chunks of rock have fallen away and more are close to falling.

There are two kinds of weathering: mechanical and chemical. In **mechanical weathering,** rock physically breaks apart. This is the main form of weathering in dry regions, such as the southwestern United States. Mechanical weathering can be caused by plants, ice, or water. Plant roots growing into a **joint,** or crack, in a rock may widen the joint and create new cracks. Ice that forms within joints and holes in a rock can shatter the rock into many smaller pieces. This is because when water freezes to form ice, it expands, or gets bigger. As the ice expands, it breaks apart the rock the way it would burst a bottle of water or can of soda pop if it froze.

In **chemical weathering,** substances in the rock go through chemical changes that weaken the rock. This is the main form of weathering in areas with humid, or damp, climates, such as the eastern United States.

Some chemical weathering occurs when rock makes contact with air. For example, iron within rock combines with oxygen in the air to form rust. This gives many rocks a reddish color. Often the rusted material breaks apart and loosens, making it more easily carried away by wind or water.

Chemical weathering also occurs when rock touches water in the ground. This happens most commonly with rock that is rich in limestone. In dry climates, limestone resists erosion, but the abundant water in humid areas makes limestone change form and dissolve away.

A great example of this process occurred in 1981. The rock beneath the town of Winter Park, Florida, is made almost entirely of limestone. Over the years, erosion started by chemical weathering weakened the limestone until it was no longer strong enough to support the soil and the buildings above it. In 1981, the ground caved in, forming a **sinkhole** that engulfed a house, five cars, and part of the city swimming pool.

The sinkhole in Winter Park, Florida, was 400 feet (122 m) wide and 125 feet (38 m) deep.

A sinkhole is one effect of erosion that seems to happen suddenly, but it is actually just the final, dramatic result of a long, slow process. A mud slide occurs after a long process, too. Several times in the 1980s and again in 1998, rain-soaked hillsides in southern California turned to mud and collapsed, burying roads, houses, cars, and animals. Although the mud slides were sudden, erosion of the soil on those hills had been going on for a long time.

Even without a sudden event, the results of erosion can be spectacular. For example, the amount of material carried off by a stream in one day is small, but over thousands or millions of years, erosion can change the landscape. It took the Colorado River 5 million years to carve its way down through a mile of rock, but the result is the Grand Canyon.

The effects of erosion can be spectacular. Bit by bit, this river in Wyoming cut straight down through more than 200 feet (61 m) of rock.

9

THE FORCES OF EROSION: WATER, GLACIERS, AND WIND

When the wind whips up a dust storm that stings our eyes, its ability to move soil is very clear. But the most powerful erosive force on earth is not wind but water, which causes erosion in its solid form—ice—and as a liquid.

Water in its liquid form causes erosion in many ways. Streams—from tiny creeks to huge rivers—carry tons of eroded earth every year. The size of the eroded material a stream can carry depends on the speed of the stream and how turbulent, or rough, the water is. A fast, turbulent stream can carry large rocks, while a slow, gentle stream carries only smaller particles of sand and clay.

Water—the most powerful erosive force on earth

Pulled down by gravity, eroded material eventually settles out of the stream when the stream no longer moves powerfully enough to carry it. A large river may carry its load for hundreds of miles. Some material drops out of the current along the river's **floodplain.** This is the flat area, on both sides of a river, that is likely to be covered by water when the river floods. At the river's mouth, the place where it joins an ocean or a large lake, the river slows down even further. There it fans out and deposits more material in a broad area called a **delta.**

Water causes erosion even before it reaches streams. A simple rainstorm loosens soil and carries it off through **sheet erosion.** The rainwater flows in a thin layer over the surface of the ground until it reaches a ditch or stream. Once in the ditch, the water erodes the channel even more, cutting deeper into the soil or rock it flows through.

Above: *This river has deposited eroded material to form a lush floodplain.* **Right:** *The rocks on the near bank of this river don't match the rock face on the far bank—a clear sign that the river carried the rocks from upstream.*

A flash flood can move a great amount of soil. As floodwater rushed across this field (above), it dumped almost 3 feet (0.91 m) of sand along this fence line. The wavy pattern in this dry, sandy streambed (right) was left by rushing water.

In arid, or very dry, regions such as the southwestern United States, many streams swell with rain and melted snow in the spring, then shrink to a trickle by July or August. After the spring rains, it often doesn't rain again until the violent storms of late summer. When fierce summer rains hit the dry soil, most of the water runs off into the nearly dry creek beds. Within minutes of a big rain, this water surges over the stream banks to form **flash floods** that can wash away roads, bridges, and cars, as well as tons of soil and rock.

Not all water from rain and melted snow flows over the surface of the land. Some water seeps down into the ground, especially where the land is flat. There it fills the joints and holes within rocks and causes chemical weathering.

Oceans and lakes also cause erosion. Particles such as dust, sand, gravel, and seashells are pushed up onto the beach by incoming waves. Later they wash back into the water on other waves. Over time, the waves create beaches that look much the same from year to year. Dunes are in the same places, and shells washed in by the surf pile up at the same spots. But the individual grains of sand that make up the beach are continually being traded. They wash up onto the beach, help form the landscape, and eventually get whisked away again, to be replaced by other grains of sand from the ocean depths.

Rocky shorelines erode differently from sandy beaches. As the endless motion and power of the water erode softer rock along the shoreline, the remaining erosion-resistant rock forms arches, caves, and chimneylike **sea stacks.**

Top right: *Sea stacks off the coast of Oregon*
Right: *The sand on a beach is constantly moved by the waves, even if the beach always looks the same.*

The icy arm of a glacier cuts through the mountains of Alaska.

Water in its frozen form also contributes to erosion in many ways. While tiny pockets of ice within a rock can split it into dozens of pieces, huge glaciers can carve out whole valleys and scrape a hilly landscape flat.

Glaciers form in areas that don't get warm enough for all the snow that falls to melt. The snow keeps piling up until it is packed so deep and heavy that it turns to ice, just as a snowball does when you squeeze it hard.

The conditions that allow glaciers to form occur in high mountains and near the earth's poles. During the Ice Ages, 15,000 to 20,000 years ago, earth's climate was about 10°F (5.5°C) cooler than it is today. More snow fell during the cold season, and much less snow melted during the warm part of the year. As a result, glaciers didn't develop only in the high mountains and polar regions. They grew until they covered much of the land that is now Canada and the northern United States, as well as northern Europe and Asia, and the southernmost parts of South America.

Glaciers are solid ice, several hundred feet to several miles thick. The tremendous weight of all that ice presses so heavily on the bottom part of the glacier that the ice there changes form. Although it still looks hard and brittle, it softens and begins to flow very slowly. At the same time, ice crystals all through the glacier gradually slip past each other. These small movements carry the glacier downhill.

A glacier moves a few inches to several yards each day. Whenever more snow gathers at its higher, colder end, the added weight helps push the glacier down the mountainside. The leading edge of the glacier is the end that is at a lower elevation where the temperature is warmer. This part of the glacier often starts to melt. If snow is added to the high end of the glacier faster than it melts from the leading edge, the glacier **advances.** It gets wider and thicker, and its leading edge moves farther downhill. In warmer periods, the glacier melts so fast that the new snow can't keep up. Then the glacier shrinks in width and depth, and **retreats** up the mountain valley it carved.

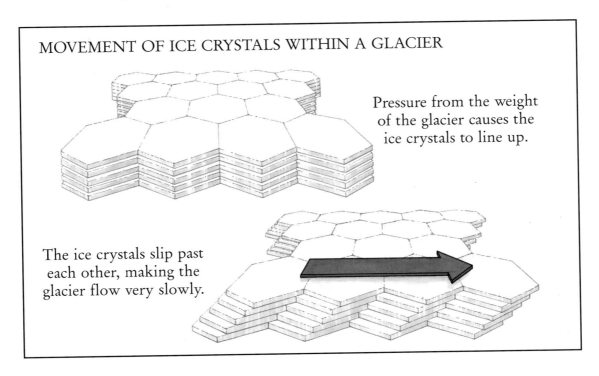

MOVEMENT OF ICE CRYSTALS WITHIN A GLACIER

Pressure from the weight of the glacier causes the ice crystals to line up.

The ice crystals slip past each other, making the glacier flow very slowly.

Because a glacier is both hard and heavy, it erodes almost everything in its path. It scours rock, scrapes away soil, and even dislodges big rocks and boulders. These in turn scrape up other rocks as the glacier creeps along. The glacier then carries these chunks of earth with it as it flows. When a glacier retreats, the soil, gravel, and rocks it has been carrying fall to the ground. Rocks that are dropped by a glacier are called **erratics.** Sometimes eroded material piles up into ridges or hills called **moraines.** The weathered rock left behind can be further eroded by streams, storms, and high winds.

This field in Maine is littered with glacial erratics.

In some places, sand carried by the wind piles up into dunes.

Although wind can't carry boulders or rocks, it can sweep away huge amounts of sand, dust, and other bits of earth. Like water, wind also drops its load of eroded earth when it slows down. It usually deposits fine particles across a wider area than a river does. If the wind slows down at about the same place day after day and year after year, the eroded material it drops may pile up to form hills or dunes.

Wind moves eroded material in different ways, depending on the size of the particles and the strength of the wind. A strong wind lifts the finest particles, like dust, high in the air. It may haul them hundreds or even thousands of miles. Slightly larger particles, such as fine sand, jump into the air a few feet and fall to earth after traveling a short distance. They may jump again as the wind continues to blow. The same size particles in a lighter wind, or slightly larger particles in a strong wind, roll and tumble along the ground. Larger particles, such as pebbles, stay on the ground but are nudged along by the wind. In these ways, wind can move tons of earth many miles.

Some of the best farmland in the world was created by wind erosion. Over millions of years, wind scooped up fine, dustlike material called **loess** (LUSS) from deserts and from glacier-scraped mountains. It then deposited layers of loess up to 200 feet (61 m) deep in large areas of the Ukraine, China, Argentina, and the Great Plains of the United States and Canada. Over time, as plants, earthworms, and microorganisms started growing in it, the loess became very fertile soil that people use to grow crops like wheat and corn.

BENEFICIAL EROSION

In many parts of the world, eroded earth carried by rivers forms nutrient-rich soil that allows people to farm in areas that would otherwise be poor for farming. For centuries, the Nile River flooded the arid lands of northern Egypt every year. With each flooding, the river deposited more than 9 million tons (8.2 million metric tons)—more than half a million dump trucks full—of eroded soil from the mountains of Ethiopia. This rich soil, left on the Nile floodplain and delta, provided excellent farmland.

In 1970, the Egyptian government finished building the Aswan High Dam across the Nile. The government wanted to save water for dry years, prevent damaging floods in wet years, and allow people to farm year-round.

But the dam doesn't just hold back water. It also catches the tons of precious soil that normally helped Egypt's farms to thrive. As a result, farms downstream from the dam no longer receive fresh, fertile soil each year. And without its yearly dose of new soil, the Nile Delta is being eroded by the waves of the Mediterranean Sea. The natural balance between addition and erosion has been destroyed.

Opposite: *Dusk settles over glacier-scraped mountains in Montana.*

Water, glaciers, and wind all carry eroded material. Eventually, pulled down by gravity, all eroded material comes to rest on the earth again. There it forms new soil or rock. It may form a river delta or a floodplain. Over millions of years, it may help form new mountains by being lifted up by an earthquake or volcano. Then the winds and waters go to work on it again, as erosion continues to whittle away at the earth.

RESISTING EROSION

Not all types of rock or soil react the same to the forces of erosion. Soft rock, like shale, usually erodes faster than hard rock, like granite. Hard rock that contains many cracks may wear away faster than softer rock, however, because the cracks can cause big chunks of the rock to break off. Soft rock will lose just tiny bits at a time.

Most soils erode more quickly than rock, since soil is made of smaller particles that are more easily carried away by wind or water. How well soil can resist erosion depends on factors such as plant life and climate.

Layers of erosion-resistant rock jut out from this rock formation. The light-colored rocks on the ground are chunks that have broken off from harder layers at the top of the formation.

Left: *Plant roots give soil something to cling to.* **Above:** *Bare soil is unprotected from the direct impact of raindrops.*

Soil is the layer of earth in which plants grow, and plants are the key to keeping soil from eroding. Plants work both above and below ground to protect the soil. Above ground, their leaves shelter the soil from direct hits by raindrops. It may sound funny to talk about the force of raindrops, but scientists have found that a single raindrop strikes the ground at 9 to 20 miles per hour (14.5–32 km/h). That's forceful enough to loosen bits of soil, which may then be washed away by more rainwater. If the raindrops strike a leaf first, they fall to earth more gently and do not disrupt the surface of the soil.

Below ground, plant roots give the soil something to stick to. If you pull up a weed, you will find soil clinging to its roots. Soil with hundreds of healthy roots growing in it resists erosion much better than soil with just a few roots.

Bare soil—soil with no plants growing in it—is the most defenseless of all. In dry areas, the soil usually has fewer plants than does soil in rainy areas, so the soil in dry areas has less to cling to when winds and water come. It also has less protection from damaging raindrops. For these reasons, soil in arid regions erodes quickly.

BADLANDS

In some arid regions, erosion begins before any soil forms. These areas lack worms and other soil organisms, so few plants grow there. They also get little rain. When rain does come, it strikes the bare ground and erosion quickly digs gullies through the fragile earth.

In the United States, these areas are called badlands, a name given to them by early French explorers because of their harsh and rugged terrain. Many people think badlands are beautiful because of their complex shapes and vivid colors. The shapes come from erosion by rainwater or melted snow. The colors come from different substances in the rock. For example, iron in the rock may turn reddish. Limestone often looks white or cream colored.

In a large area of badlands in central Wyoming, red and white layers of rock form a "candy-stripe" pattern. Nearby, green and purple rocks give the badlands a very different appearance. Other large badlands are found throughout the western United States. One such area in South Dakota has been made into Badlands National Park.

The Rocky Mountains are young mountains with high, rugged peaks.

SHAPING THE EARTH

Erosion by glaciers, water, and wind has shaped the earth as we know it. Mountains, for example, are created by earthquakes, volcanoes, and processes that occur deep within the earth, but their size and shape depend on erosion. "Young" mountains tend to have sharp peaks and steep valleys. Over time—millions of years—erosion makes mountains smaller and smoother.

The Rocky Mountains of western North America are a good example of a "young" mountain range. The Rockies arose between 85 and 65 million years ago. Their highest peaks reach over 14,000 feet (4,267 m) high, and many others top 10,000 feet (3,048 m). The peaks are rocky and jagged. In some places, glaciers still grind away at the mountainsides.

In eastern North America, the Appalachians are an example of an "old" mountain range. They formed over 250 million years ago. They were similar to the Rockies back then, but erosion by glaciers, water, and wind has worn their highest peaks down to less than 7,000 feet (2,134 m) high. Their peaks are rounded and are no longer high enough to be cold enough to have glaciers.

The Appalachian Mountains are worn down from more than 250 million years of erosion.

A glacier scooped out the left side of this mountain in the Canadian Rockies.

In high mountain regions around the world, glaciers shape valleys and help form lakes and streams. Glaciers dig deep into the sides of mountains, forming valleys that may be round or long and narrow. As glaciers retreat, water from rain and the melting glacier often fills these deep valleys to create lakes. The Great Lakes of the United States and Canada are glacial lakes. So is Loch Ness in Scotland, home of the mythical Loch Ness monster. Loch Ness is over 24 miles (38.6 km) long, just 1 mile (1.6 km) wide in most places, and over 750 feet (228.6 m) deep.

V-shaped valleys such as this are often formed by rough streams that flow down steep mountainsides.

Not all water from mountain glaciers creates lakes. Some pours down mountainsides, carving streambeds and carrying away much of the loose material scraped up by the glaciers. Some of this eroded material is very large. Rocks and even boulders that started out on a mountaintop can end up in a distant valley.

All rivers and streams continually eat away at the earth, shaping valleys and other features. On steep mountainsides, the swift, rough streams cut deeply into the earth, creating V-shaped valleys. When they reach a more level area at the base of the mountains, the streams drop much of the eroded material they've been carrying. This material forms broad, gentle slopes where the mountains meet the plains.

Slower streams often don't cut as deeply into the earth as swift mountain streams. Instead, they may meander, curving first one way, then another. A valley formed by a **meandering stream** is usually broader and more U-shaped than one formed by a swift-moving stream.

Above: *Some meandering streams don't form broad valleys. This stream cuts a narrow path through the rock at the lower right of the photo (1), enters a small valley (2), then winds through steep cliffs again at the upper left (3).* **Right:** *A meandering stream makes sharp curves through a pasture.*

A braided stream

Sometimes, especially in springtime, a stream will spill out of its shallow bed and make new channels that intersect, or cross, each other. This forms a **braided stream.** A stream like this can form a very fertile floodplain by depositing eroded material over the wide area covered by all its channels.

In areas with rock that resists erosion more, most streams don't meander or form braids. They stay in one narrow channel, cutting way down into the rock to form a **canyon,** like the Grand Canyon. A canyon is a deep, narrow valley with very steep or even vertical (straight up-and-down) sides.

Sometimes the higher land that separates two river valleys gets pinched off by further erosion so that it stands out like a flat-topped mountain. This is how a **mesa** forms. *Mesa* is a Spanish word that means "table." The top of a mesa is called the **caprock.** It is made of very hard, erosion-resistant rock that wears away more slowly than the layers of softer rock underneath it. As water from rain and melting snow flows down the sides of the mesa, the softer rock erodes, leaving some of the caprock hanging over empty space. It may then break off, reducing the size of the mesa.

Right: *The white boulders on the ground were once part of this mesa's caprock.* **Below:** *Grand Mesa in western Colorado is one of the largest mesas in the world.*

Erosion by water and wind continues to eat away at the mesa until it becomes a **butte,** which is just a smaller version of a mesa. Eventually, the butte too shrinks until just a tall, thin **pillar** is left. In Bryce Canyon, Utah, hundreds of pillars stand side by side. Geologists, or scientists who study the earth, think the pillars probably formed from a mesa that had hundreds of joints in it, so that water and wind eroded all the rock that once filled the spaces between the pillars.

The pillars in Bryce Canyon, Utah

A LOOK THROUGH TIME

As a river erodes the rock beneath it, different layers of the rock become visible. In Arizona's Grand Canyon, the whole story of the Colorado River's erosion of the landscape can be read in the rocks of the canyon walls.

Most of the layers are made of sandstone, limestone, or shale. These layers were deposited many millions of years ago by streams, ancient oceans, or wind. Older layers were covered by younger layers. The layers vary in color because of the different chemical substances they contain. They also contain different kinds of fossils, or remains of living things that have turned to stone, depending on what kinds of plants and animals were alive when the layers were deposited.

The river started carving the canyon about 5 million years ago. Some of the layers, like those of limestone and sandstone, were very resistant to erosion. As the river cut down through them, they formed cliffs that were almost straight up and down. Other layers, like those of shale, were softer. Here, the river eroded the sides as well as the bottom of the canyon. Instead of steep cliffs, these layers formed gentler slopes.

This weird-shaped rock was sculpted by wind that was carrying sand.

Compared to glaciers and rivers carving out whole valleys, and water sculpting canyons, wind's effects on the landscape seem small. Wind that's carrying sand can shape rocks and mountains, just as rubbing sandpaper on a block of wood can round off its edges. But since wind rarely lifts sand higher than 6 feet (1.8 m), it works mainly on rocks close to ground level. In some windy areas, boulders look like mushrooms or light bulbs, with a narrow stem at the bottom and a larger dome on top. Sand carried by the wind has eroded the bottom part of these boulders while not reaching high enough to wear away their tops. Wind also creates **arches** by eroding the center portions of stone slabs that stick up out of the ground. Hundreds of arches have developed from sandstone slabs in the southwestern United States.

One of the most unusual landforms created by wind erosion is a **bowl.** This is a place where strong winds scoop away the earth for thousands of years until a huge, bowl-like depression develops. Bowls can be found in Africa, Asia, Australia, and the western United States. One large bowl near Laramie, Wyoming, called Big Hollow, is 9 miles (14.5 km) long, 2 miles (3.2 km) wide, and up to 120 feet (36.6 m) deep.

Above: *Landscape Arch.* Below: *South Padre Island, Texas, is a barrier island with rows of hotels and beach houses.*

Along the seashore in some parts of the world, wind and water erosion combine to form **barrier islands.** These are islands made of sand that has been deposited by the wind and waves. Since they have nice sandy beaches, many barrier islands, such as the Sea Islands of Georgia and the Outer Banks of North Carolina, are popular resorts where people go to swim, boat, and fish. But barrier islands are constantly shifting because the wind and waves erode the beaches as well as add new sand to them. Erosion created the barrier islands, but it also makes them poor places for building large houses, hotels, and towns.

DESTRUCTIVE EROSION

On a barrier island in North Carolina, a luxury hotel crumbles into the ocean. In California, a mud slide engulfs dozens of homes and buries a highway. In Iowa, heavy rain sweeps tons of rich soil from a farmer's fields.

These examples have two things in common. They show how erosion can destroy things of value to humans. And in each case, the erosion became more destructive because of something humans did.

Rainwater drains from a flooded farm field, taking loads of soil with it.

Barrier islands along the southeastern coast of the United States became popular vacation spots in the early 1900s. More and more people built houses on the beach. People also built walls between their houses and the water to protect the houses from erosion. But these walls have actually made erosion of the beach worse. Before the walls went up, wind and waves removed sand but also brought in new sand. The walls allow sand to wash away from beneath them but block new sand from being deposited on the beach. Waves can then eat deeper and deeper into the islands, eventually scooping out so much sand that the walls—and the buildings behind them—collapse. Whole beaches have shrunk, and the plants and animals that lived there have lost their homes.

A house collapses into the ocean due to erosion of the beach on the Outer Banks, a barrier island in North Carolina.

This apartment building in Ventura, California, was crushed by a mud slide in February 1998.

The mud slides in California happen largely because people change the landscape so that it erodes too quickly. People living there cut down many of the native trees and bushes so they can build more houses and have clear views of the ocean. Without those plants, the soil has less protection from rain and fewer roots to cling to underground. The weight of the houses puts an extra burden on the steep hillsides. When the area receives a lot more rain than usual, the soil turns to mud and becomes unable to hold itself up. Tons of mud rumble down the hillsides, crushing the houses as if they were made of popsicle sticks.

Erosion of **topsoil** from farmland has become a serious problem in many areas around the globe. Topsoil is the upper, most fertile layer of soil, the layer our crop plants need in order to grow well. In the 1700s, the layer of topsoil in the Great Plains of North America was about 9 inches (23 cm) deep. By the 1990s, it was only about 6 inches (15 cm) deep. U.S. farms lose an average of over 4 tons (3.6 metric tons) of topsoil from every acre every year.

THE DUST BOWL

Not all soil erosion is due to water. Wind also carries off huge amounts of topsoil, especially in arid regions or in normally wet areas that are suffering a **drought,** or an extended period with no rain.

Wind caused some of the worst soil erosion ever seen in North America. In the 1930s, following several years of drought, high winds across the Great Plains swept up huge clouds of topsoil. Because of deep plowing, the soil was very fragile. Because of the drought, crop plants struggled to survive and were unable to hold the dry, dusty soil in place.

Scientists estimate that in May 1934 alone, over 300 million tons (272 million metric tons) of topsoil were lost over a vast area from Texas to Manitoba and from Colorado to Kentucky. And the wind didn't carry the soil just to a neighboring county or state. It lifted the fine dirt high in the air and carried it hundreds of miles. During one especially windy period, people in Washington, D.C., saw huge clouds of swirling dust that had come from farms in Kansas and Oklahoma, the center of the "Dust Bowl."

The loss of soil made farming impossible and forced thousands of people to leave the plains. Many of these people went west, especially to California. When the winds slowed and the rains returned, the land recovered enough for crops to grow. The Great Plains again became one of the world's chief wheat-growing regions.

Could something like the Dust Bowl happen again? We have improved our methods of farming and preventing erosion, but our farmlands still lose huge amounts of topsoil each year. If drought and high winds returned to the area, we could again see our rich topsoil, and the hopes of farmers, blowing away on the wind.

Planting crops along the contours of hilly areas helps prevent rainwater from flowing downhill and carving out a gully.

Most loss of topsoil is due to what humans have done with the land. In the 1800s and early 1900s, many farmers plowed straight up and down hillsides. The plowed furrows provided little gullies for rainwater to run into. This sped up erosion by making it easier for the water to create bigger gullies. To avoid this, most U.S. farmers have switched to **contour plowing.** That means they plow in curved lines along the contours or outlines of the hills. Each row of plants makes a terrace, or small level area, that can catch water. This leaves no quick, easy way for water to run down the hill and carve out a gully.

Contour plowing has helped preserve the soil, but many other problems remain. Plowing of any kind loosens the soil and breaks it into smaller pieces. These small pieces are easily washed or blown away. Growing just one kind of plant in the same field year after year hurts the soil, too. The earthworms, tiny plants, fungi, and microorganisms that help protect soil from eroding have a hard time surviving in soil that always has the same kind of crop growing in it. This is because each kind of plant takes certain nutrients from the soil and provides other nutrients in return. If a field has only one kind of crop year after year, its soil becomes rich in a few nutrients and poor in all others. Crop rotation, or growing a different crop in a field each year, helps prevent this problem and keep the soil organisms healthy.

Erosion is also a problem in many forests that are used as a source of lumber. Where the woods are **clear-cut**—that is, where every tree in an area is cut down—the soil is left bare, with no protection from wind and rain. When clear-cutting is done on steep hillsides, the soil has little chance to stay in place. With the first big rain, or the melting of snow in the spring, soil is washed into gullies and streams.

Road building, new housing developments, mining—anything that removes plants and disturbs the soil—can lead to destructive erosion. Even things that people do for fun can cause problems. Mountain bicycles and off-road vehicles often harm the soil, killing plants, digging ruts, and speeding up the rate of erosion.

Patches of this hillside have been clear-cut by loggers.

PREVENTING THE DAMAGE

Some destructive erosion will continue to happen simply because rain will continue to fall and wind will continue to blow. But much of it can be avoided by changing the way we do things.

Unless we are willing to see buildings on barrier island beaches collapse into the ocean, and the beaches lost as well, we shouldn't put houses and hotels near the water. Trying to protect them from the wind and waves only makes the situation worse.

Riverbanks can sometimes be strengthened against erosion. This riverbank is protected by riprap, a layer of rocks held in place by chain-link fencing. The water is at low level; at other times it comes up as far as the riprap.

Left: *Native flowers, shrubs, and grasses protect the soil on this hillside from erosion.* Right: *This row of trees forms a shelterbelt.*

People are realizing that they can prevent erosion by taking better care of plants and soils. Some loggers use **selective cutting,** where they remove just a few trees from a forest, rather than clear-cutting. Many homeowners leave the native plants in place around their new houses. Such practices are especially important in hilly areas. Planting grasses and shrubs on bare hillsides near houses and along roads and streams helps keep the soil anchored so it does not erode so easily.

Farming methods are changing, too. **Shelterbelts** of tall shrubs and trees between farm fields protect the soil in the fields by blocking the wind. Late in the summer, some farmers plant special crops that stay in the ground over the winter. Their roots and stems help anchor the soil and keep it from washing away in the spring rains. Crop rotation helps keep earthworms and other organisms in the soil healthy.

Some farmers are even giving up their plows. Instead of using a plow to dig into the soil and turn it over before planting, they plant their seeds in little holes punched in the soil by a special machine. This greatly reduces erosion. It also lowers the number of weeds, because a plow buries weed seeds underground where they can sprout and grow. Without plowing, most weed seeds stay above ground where they cannot sprout. And since this method of planting doesn't disturb the earthworms and other organisms that live in the soil, those creatures thrive. They in turn help the crops grow better.

THE MIGHTY EARTHWORMS

Earthworms are good for gardens and good for the soil. Farmers, gardeners, and scientists have known that for more than a hundred years. We are also learning that earthworms can greatly slow down soil erosion.

It works like this. Healthy soil is home to huge numbers of earthworms—as many as 600,000 worms per acre (1,480,000 per hectare). Each worm burrows through the soil to reach the surface, where the worm collects dead leaves it pulls into its burrow and eats. Earthworm burrows may extend as deep as 8.8 feet (2.7 m) below the surface, and each earthworm burrow is $\frac{1}{8}$ to $\frac{1}{2}$ inch (3–13 mm) across. Healthy soil with lots of worms in it has about 145 burrows per square yard (per 0.9 sq m) of land.

What do earthworm burrows have to do with erosion? They provide drain holes where rainwater can run down into the deep layers of soil. This allows more rain to soak in rather than running off the surface and carrying eroded soil with it.

EROSION: A CONTINUING PROCESS

Erosion is going on all around us. Some erosion, like the formation of gullies on bare slopes, is easy to see. Other erosion happens so slowly that we will not notice it during our lifetimes. Mountains age. Rivers cut deeper into the earth. Wind polishes the boulders of the American West.

With the forces of erosion constantly eating away at the earth, why isn't everything flat? Will the sea cliffs eventually all be worn down, the pillars and arches dissolved into sandy hills, the Appalachian and Rocky Mountains leveled like the Great Plains?

Erosion goes on all the time, but it is not the only force acting on the earth. Forces deep within the earth, like volcanoes and earthquakes, are continually building new mountains and cliffs, even while erosion is wearing others down.

Yes, the landforms that are familiar to us—the arches, the sea cliffs, even the great mountain ranges—will probably be gone someday. But by the time that happens, millions of years in the future, other mountains and rocky shores will have formed. And the process of erosion will continue to work on them as it has worked for millions of years on the features of our world.

GLOSSARY

advance: forward or downhill movement of a glacier's leading edge

arch: a rock formation that results from wind eroding the center of a curved slab of rock

barrier island: a long, narrow island along a seashore, made of sand and gravel dropped by wind and waves

bowl: a large depression in the landscape created by wind erosion

braided stream: a stream that flows through shallow, intersecting channels

butte: a flat-topped landform that results from erosion of a mesa

canyon: a deep valley with steep sides

caprock: the top, erosion-resistant layer of rock on many landforms

chemical weathering: the weakening of rock by chemical changes

clear-cutting: cutting down every tree in an area

contour plowing: plowing along the outlines of hills to prevent soil erosion by water

delta: a broad area where a river flows into an ocean or a large lake

deposit: the act of wind or water dropping the eroded material it has been carrying

drought: a long period of little or no rain

erosion: the movement of rock, soil, and other bits of earth due to water, wind, or glaciers

erratics: rocks or boulders moved by glaciers

flash flood: a sudden, powerful flood that occurs during or after a rainstorm, usually in an arid area such as a desert

floodplain: the low, flat area beside a river that is covered with water when the river floods

glacier: a huge mass of moving ice

joint: a crack in rock

loess: deposits of very fine, windblown dust that can form rich farmland

meandering stream: a stream that follows a twisting, winding path

mechanical weathering: the physical breakdown of rock

mesa: a large, flat-topped landform

moraine: a hill or ridge made of material dropped by glaciers

pillar: a tall, thin rock formation

retreat: the backward or uphill movement of a glacier's leading edge due to melting

sea stack: a rock formation formed by wave and wind erosion along seashores

selective cutting: cutting down some trees in an area but leaving most to continue growing

sheet erosion: erosion of soil by water flowing in a thin layer over the surface of the land

shelterbelt: trees and shrubs planted around and between farm fields to reduce soil erosion by wind

sinkhole: a pit where the ground has caved in because the rock below it was weakened by erosion

topsoil: the upper, most nutrient-rich layer of soil

weathering: the weakening or breaking apart of rock; the first stage in erosion

INDEX

Additional photographs are reproduced through the courtesy of: © Norman O. Tomalin/Bruce Coleman, Inc., p. 6 (left); Tom Stack and Associates: (© Milton Rand) p. 6 (right), (© M. Timothy O'Keefe) p. 8, (© Bob Pool) p. 13 (top), (© Matt Bradley) p. 33 (bottom); © Noella Ballenger, p. 14; Ecostock: (© Lee F. Snyder) p. 16, (© Ed Darack) pp. 17, 28, (© Robin Cole) p. 24; © Richard Cummins, p. 30; © Diane C. Lyell, p. 31; USDA, p. 34; Prof. Orrin H. Pilkey, Jr./Duke University Department of Geology, p. 35; AP/Wide World Photos, p. 36; Library of Congress, p. 37; © Harriet Vander Meer, p. 38. Illustration on p. 15 by Laura Westlund, © Carolrhoda Books, Inc.

J. David Love

ABOUT THE AUTHOR

As a child growing up in Utah, author/photographer **Cherie Winner** took many family trips through the spectacular erosion-shaped landscapes of the desert Southwest. Since then, she has spent many years in the Appalachians and the Midwest, where she learned first-hand about limestone weathering, "old" mountains, and loss of topsoil. She eventually settled in windy Wyoming with her dog, Sheba, and her cat, Tucker. Dr. Winner holds a Ph.D. in zoology from Ohio State University. She is the author of four Carolrhoda Nature Watch titles, *Coyotes, Salamanders, The Sunflower Family,* and *Trout.*